I0463433

SHE

IN THE LOVE OF WORDS

ELEONOR ELIZABETH ESCAURIZA HEMPEL

Copyright © 2012 Author Name

All rights reserved.

ISBN:
ISBN-13:

DEDICATION

To:

God whom has been helping and loving me, in all my ways, from my mother's womb, thanks Him I am alive and happy. Writing his mercy every day because He has blessed my ways with His love, patience, and grace.

CONTENTS

ACKNOWLEDGMENTS

The life   and   the wisdom of God, which   drive us to a better life and love.

To my mother, who has been an example of strength and patience.

To my father, whom   has   given me   words of natural good vision of the truth values, with a good heart, even though he has not an easy life.

To my children whom are the love of my lives, with whom I have grown in the knowledge of love and grace of God.

"Your word is a lamp for my feet, a light on my path"
  Psalm:119:105

1 SHE AND HER FAMILY

She was the fifth girl of a large family. She was a happy natural girl living in a complicated world between the human and divine words and world.

Her grand grandfathers was born in Europe, Her grandfathers and her fathers were born in South America.

Her grandmother was born in Paraguay, in the country side. Being a good mothers and wife.

Her fathers was not rich, but they have the mind of manufacturing and create things for a living. They have the blessings of open doors for merchandising.

Her grand grandfather, from her mother, was scientist working in South America, Argentina.

Her grandfather was a men who likes to work in the production and exportation of leather, before the world war two.

After the war, He started to work with wood production and exportation, being one of the first in South America.

Her grand grandfather from her father comes from Spain and Portugal.

Her grandfather was born in Paraguay, working with the sugarcane and agriculture in his country. Her grandmother from her father was a beautiful and sweet Paraguayan woman.

She's father worked in textile and button trade.

Her parents started a good life together.

The family was good in the standard level of living.

When the family grow, the economic system was in trouble in the country and the families as well.

She, was growing looking the world around.

Having her eyes and her ears in the word and world of people.

She has light brown hair. A beautiful face and smile.

She was not tall, but not too short.

Her skin was white with glowing complexion.

At school, she was smart and she likes to play with their classmate.

She was very good in exclamation and writer.

She was not rich, their parents has to find a school for her and her brothers and sisters, for many years during their school age.

Her father has been working in the complicated world of business with the competition of the time.

Many times he was out of home, trying to sale his products,

in the way to bring the food on the table.

Many times he was discourage and went to their friends for a drink.

These situation created trouble in the family, but the father mind of creation and sale was in her father work. Her mother has been cooking and sowing, she was a strong example to make a better place to live, it was not easy, but it wasn't the end.

She listen the words that they speak, trying to understand the real situation that she was in, according to other people and according the true that she was living.

She has the heart to help and work with her mother.

Her older sister help her in her homework and the daily bath, and cleaning the house.

Creating things for a living, cleaning, cooking, and playing all the time with her sisters and her brothers, was her style life from that time.

We can say that they were poor, but with the will to advance for a better life and good future, but there were many doors closed.

Her fathers have many trouble in between.

She was very aware of the situation, trying to help her family.

She had been brought good reports from school, until a

school system problem make her failed.

2 CHILDHOOD

There was the time of favoritism. Teachers were called to make some kind of difference with the classmate. Was the Political Dictatorial time and religious system. Her father was not with the political system of the time, making this situation be worst.

She has been growing in this environment, hoping that the situation can be changed.

Her family have to move from their house many times, renting houses.

She liked it because she could have new friends to play, but she wanted to be with the older friends also.

In the world of word, she was searching something more, that can guide her to a better life.

She was looking for the divine side of the world.

Her parents didn't go to church. But they have a bible at home.

She started to learn the words and learn how to know God.

She was searching God through the nature, plants, birds, weather, news, colors, and words.

In the world of human and divine words.

It was not easy because people lived their lives, she has her family European and South American educational system at home.

The food, the clothes, the point of view, was a little different.

The place where she was living was not expensive, they can have the fruit, vegetables, cereal for not much.

The have patio plenty of trees of fruit in her house.

Her mother knows how to make European and South American plates- sometimes you just have to change the name of the recipe.

Food as patty, cakes, bread and jams fruit.

Her mother even cooked for sales, as her big brother and sister also.

She help them happily with all of these, because she will have the permission to weak up early in the morning to see the dawn.

One day her mother failed sick, her heart was not well, and she have to move to the country side.

In that place she can find the school, friends, places to play until they were so tired and go to sleep.

The house was big with a big back yard.

The mother get better, but for the father was difficult to

stay at home.

3 AT SCHOOL

She likes school very much.

She loved the teachers very much. The classmates, boys and girls, were always willing to play.

In that time were many kind of plays to share outside the room.

She was calling for their teachers for declamation for the festival days.

The hours at school went fast for her. She took that time to expand her world.

They share the sit with the classmate also.

The break time was for playing and fight at the cantina place because there has always full, and she was not that tall enough.

Many times she brought the food from home.

 She was the first in her class, until almost finish the school age. She was interrupted because the favoritism in the political, social and religious system.

When she was in the line, waiting for her diploma of honor, as years before, she is communicated that she hasn't approved the school year.

She didn't understand the situation, she ran to her father showing the report card. Her father wrote a letter to the teacher and to the principal of the school, make them notice that the report card was erased, the qualification has been changed.

Without having resolved the situation she left the school.

In that time, a person can called to interrupt a children future, because they have someone else. It was the dictatorship time. The father of she was not in the political agreement, and he didn't kept that in silent.

Having passed many years, she started again the school and finished.

She finished in the first place again, with the diploma of honor again.

After that, she decided to get in the high school, where she can make two level in one year, in the way to recuperated the years that she has been lost.

It was a good idea, until she had problem with the papers, because she need three month more for the High School legal age.

She started to work, looking for a school again.

She went to school after finished the work.

In between, she got married and she has a daughter.

She finished the school when she was pregnant.

Marriage

She got married with a men with good surname, graduate in Administration, he was a little older than her.

She keep studying, but knowing that she has two children.

After a time, she get in the Economy University. She was learning very fast about account, economy and legal right.

But, her husband get sick and she has to left the university.

She got at home, dedicating the time to their children.

She had to learn high couture, because the propaganda said it was for people who doesn't sew any button. She was in that situation.

She took the couture class to make their children clothes, than she make clothes for people in the way to earn money for their children school.

There was time where she wasn't good and she has to go to the doctor for medicine for a while.

She has got stress. The marriage was not with good communication.

She has been trying to make him involve in the same communication, writing, taping praying, learning, talking and teaching to their children.

In between, she went for a title of Executive Secretary at CAES, an Institute of the National University, in Paraguay

4 WORKING IN THE FIELD

Then she started to work with the missionaries who came to work in her country.

She was the secretary and took the biblical courses to be a biblical instructor.

It was a good time, knowing more about God.

She started to learn English. Her life started to change.

Studying the word of God, she knows more about the light of life, she finally find that she has been looking at, God's original character and His coherent style of life.

She started to teach with the missionaries and teachers who came to work to her country.

Her marriage was without true communication, the compromises of the world, different world and friends, the education of the children, was difficult for them to have a communication with right balance in life.

It was not religion, it was different world and the ignorance of where the boat is headed, without true communication.

She learns Languages

After these, with four children, she need to left the work in the way to educate and take care of them, closer.

Having differences in the marriage, without true communication, she gets divorce, even though they were trying twenty years.

With the health a little cracked, the doctors give her some medicine for the stress and stomach to recover again.

She was working teaching Spanish to English people in an American school from Asunción.

She keeps learning languages and started the Business Administration career.

She started the career working with the national authorities from her countries, in the exposition of handicraft and artisan textile.

She worked with the Paraguayan artisan in the innovation products and home decorating products.

She went to a European seminary for artisans in her country.

There where Spanish, France, Paraguayan designers and Producers in the seminary.

She learn a lot from the teachers and friends.

She has relation with a men, when she realize the word of God when it said: will two walk together if they are not the same?

She separated from the relation, trying not to make the same mistake, and she started to dedicated to the word of

God in all her ways asking Him be with her every single day of her life.

After these situation, she open a webpage to expound and sale the paraguayan handicraft and yerba mate through internet with her sons.

She sent the products to State.

At the same time, She worked with translation, teaching Spanish and English and started to write books.

Eleonor Elizabeth Escauriza Hempel

5 BEING A MOTHER

She got at home, dedicating the time to their children.

She loves her children with all heart, doing everything in the way that they have a better life than her.

Their children was the principal reason of her life.

She has been left everything, to give them her time and be with them when they need her.

She has always working and studying in the hours that they children was a school or University.

They were growing up in the environment of love and comprehension, having differences between the word of God and what people do with.

They have to decide for the values way. Because people talks very much about God, but they disappointed in many ways.

In this time of our lives, we can see people talking, singing, sharing the word of God, but in the cross line they decided according to the preferences that they have in their heart, these is our process of knowledgement.

She prays that God give them wise and Godly Spiritual discernment.

But the life will teach us the truth through the daily life

and the word of God.

We are living in a professional and technology world. In the middle of this, our children are learning about talk and do from the heart or from the mind.

The truth is something that everyone created in their world, according to their preferences, economics, beatitudes, and many other ways as human can see and judge.

God is a different history. Fallowing God is to fix the eyes and the heart in His word without doubt. Then, we can see the truth.

If our life is not reaching the goal that we have in heart, it is not God's fault or failure, it is humankind selfish heart.

We are living in a society with many cultures, religion, education and values.

People decided according to their own understanding, without God, most the time.

She pray for her children, that people who are around their children don't take their faith from the heart as the devil does.

They will have the time of growing as we had, but God's promise is that He will help the people who we care the most.

Educate and raising is not for everyone. Children need love

and time. We learn with them to be parents. God is always there to give us a hand.

If we see our lives trying to shape them in our experiences. lacking God´s vision for our children, we can be in dangerous way.

Every humankind, born with a purpose and gift in life.

We have to be aware do not be in the middle of our children destiny, taking off God´s purpose in their lives.

Our compromise or rule in life is to love them, in the way that we can give them the best thing that we have, and wait what will the Lord accomplished in them.

In these time our best resource is to pray. Listen what they have to said or share, and be with them and for them, in the moment what they need.

It is not easy to be parents, it takes time, the best of us, it takes our lives.

6 TEACHING IN THE FIELD

She will take care three children during the mission. There she will be care with a little amount that she will send to her children for help.

She took some handicraft and yerba mate to share with the family and friends to make them know about her country.

The place was a little town like the series of movie.

It was her first time in contact with the snow.

She visit some places, friends, church where she learnt many things about the culture.

In that place God touched her heart to start to write and translate again, in the way to complete her professional and divine career.

When, she return to her country she get the enrollment to the University course of Translation, in the way to be ready for the Translation Test Registration.

In the middle of these, she started to write fast reading books in Spanish and English, with divines and human topics of values of life, in the way to understand and apply in daily life.

She continue with the handicraft and yerba mate business.

She makes home decoration home articles and cakes.

She has understood that a mission is a travel of life.

Even though we desire to help people, will be good to fallow God's advice when he said that the first thing that people has to do in their lives, is to seek Him firstly, after that, we can help them in the way they need.

She thanks God for every experience that she has in life as a missionary when she is outside the country and when she in in her country also.

To be in mission means it is a travel of daily life.

Where ever we are is the place and time that we have to have a mission and share the gifts and talent that we have.

God will guide us in our lives.

The Holy Spirit will guide us to people with words and peace.

We have the first mission to take care our families, our society, our country in the place that we live.

If we are not doing everything right, God will see our heart and He will accomplished the mission that we have in life, driven us to a better knowledgement through His word, love, justice and peace.

7 WORK AND PROFESSION

She has taken the translation test, but she didn't approve. They said that she will need more practicing in translation system.

She understood all these, and she kept practicing through internet and she got her enrollment in the University translation course, to obtain the professional registration.

She took the test, and she failed again. They said that she need more practicing again.

She decided to keep writing and keep working with the handicrafts.

She took the translation test four more times, but the University didn't get her translator registration.

With different examiner from the Union Nation who asked her to talk about the weather, climatic, eco system and social red.

With dictionaries with not all technique words, she took the fourth time test.

During the test she felt as her brain and stomach has been scammed, as the fifth time before.

At the University where the Dean has not speaks English, where the principal of the Institute of Languages was confronting many cases likes She has , with the examiner table test is acquiesced with different kind of people, with different education, political and religion, the translator registration was impossible to get.

She confront the situation to the Dean and Principal Language Institute.

She wrote a document to the University at the Language Institute, explaining them the real situation of the people who need a professional registration to work. With the truth between, where the examiner asked for a test, and how a person can take the test with the materials and time that they are given to them.

She took to them job references with documents from works that she has been doing with people who speak English as first language.

She took also the test that she has given with the mistakes of the examiner in the process of evaluation according to the dictionaries and technical words in the topics evaluation.

The university started to confront many issues with the students. They changed the authorities.

She is waiting the last resolution of her registration.

8 AUTHOR BOOKS

She have a time when she can look her live her children has growing up and is the time that She have to have the activities that She has been postponed in the childhood time of her children, in the way that She can be there for them until they reach the mature age.

She had asked to the Lord through prayers to accomplished the things that she has left before.

She has been written books into Spanish and English language, inside the environment and troubles that people confront daily.

These books has written to improve the style life from divine perspective, in the way people can understood and practice the word of God in the way that we can have a better life, beyond colors, religion, and nationality.

The topics of the books has base on Love, Divine Identity, Public Speaking, With All Heart, In Your Presence, Praying to A Living God, these last mentioned, are books for children for them integral education.

All experiences that she has been living, was embody in her heart in the way it can flow to others, through de books.

In the world, as in Paraguay also, as important nation with other nations, which have the word of God based on their laws and values, have the potential to actualize and active the word of God as principal source of wisdom and development.

People talk too much about education, didactic materials, educational instrument, meanwhile God has bequeath His word in the way to reach the individual and group potential from the childhood until to reach individual gift, and it be performed in talents and be blessed people to their society in this world, that we are living nowadays, bringing a piece of heaven on the earth.

Nowadays, we have internet, which help us to be around the world just in a second.

We have the blessings to be communicated and be able to teach and be instructed according to the word and wisdom.

We can extended our world and know more about what is going on around.

Knowing that her books has been reading around the world, is such a blessing already, and it is a way to develop and reach the gifts and help people and the way they need.

The written books through the words are living experiences exposed with wise words.

She hopes that God bless the books and its be divine instrument that guide many people applying in their lives, having a style life with wisdom, love and peace

9 TRANSLATOR AND INTERPRETER

She has decided that she will write and translate according to the God's gifts and purpose, because she knows the valuable of the diploma on the earth, but she also knows the valuable of the God's gift on the earth and heaven also.

She has been translating books will bless people around the world, changing their lives, bringing them peace, happiness, health.

The values that people can't buy, but without them, people have an empty lives, the reason for they are looking for other things and it can be dangerous things to make them happy and valuable.

She thanked God for internet, because it allows her to translate around the world.

She loves to write and translate to people who need the values of God in their life, beyond religious, color, race, status etc.

She understood that we have the life that our fathers gave us, but we can change with the word of God the things that is not right in our lives.

God is a God for everyone, because He loves us. As a good father He wants that everybody knows Him to have a better and complete life here, on the earth, bringing a peace of heaven at their home.

She has been writing many books, with topics that people can applied in their life, as simple as it is when the heart decide to change for good.

10 SHE AND HER LOVE
FOR THE WORD OF GOD

She is a person like other women of the world. With the same characteristic of the women of her country.

Inside of a society that measure the values according to the social and economic status.

She dedicated her life to know more about God´s word in the way to find a way to guide her and her family from the present and a better future.

Having born in a family with paraguayan and foreign traditions and manners, living in a country with traditional culture, with many foreign people inside of the cultural system. The word of God help her to have different point of view.

It is not just a religious environment, but as life style.

Having born with an important surname, but with small resources, show us to see as the society involve us according to their personal interest.

Meanwhile the word of God involve us to a true social, economic, and social development, toward to the divine original design for every person and society.

These divine design make her to live walk, develop and write the divine values toward to a fact that shows as true and real. It is not something new, but we have to be careful to take as real and original fact.

If we see what the human being can do, from the point of view of the social, politics and religious ego, we can lose the fact of the divine development according to the original and true resources for each person, society and nation.

Inside of a country potentially good, with fertile soil, suitable climatic for the agriculture, floriculture, cattle raising, being one the most import in south America, with a communication system which goes from the paraguayan nation to every corner of the world. Industries in many areas, and it is still name a undeveloped country.

Most the habitants in Paraguay are under eighteen years old, they are looking for opportunity of education and work, with a better future, with many politics parties and religion, where the word of God can be confused without clear direction, for people who have the good ambitions of an integral development of their people.

The word of God is light for the confused people. It is the base the all science and good personnel and social order.

The divine true, make her and her family to go the future, inside the work and education as a daily frame.

The development of their children make her established on the divine values for the management or vision and life that she can have.

She understood that having born in a family with small

resources can't be an obstacle that stop people for the integral development or be obstructed the future vision for her, her family and her people.

She, with her works and education, with the human and divine values, which has helped her though the books and translations, and interpretations that she could make, taking the light, order, peace, and strength of God where people need.

She give thanks to the technology that has been helped her with the human and divine knowledges, given her word that teach her the original and real truth, more than the education that she has been receiving, based on the word that help people to be healed and straightened the weak areas, creating opportunities in the middle of local and foreign threatened.

She believed that divine prosperity, keeps order, heath, peace, respect, education and make people grow, as a society and finally the nation.

Not just to grow but to be established leaving an integral inheritance in life.

As humankind we have the opportunity to choose the way.

It is for smart people, to decide from early age, or since we have understanding, to see the importance in taking the word of life, and not leaving room to the ignorance between the good and evil, and its consequences. Being part of a style of life which makes obstacles in our lives, and in a worst way, blame to God for our failures.

Each of us have had a style of education, which are in the main and dwelling in that place, given us the manners and actions according with what we have been receiving. But when the light comes, make our understanding be illuminated through the word of God and as result we can see the difference between the good and the evil.

Most the time we believe that to be religious person help us to be right in front of God, but God desires is that we be right in him, in the way that He can guide us and take from us the ignorance that made our lives be complicated.

She understand that the word of God isn't synonymous of human perfection, but it is divine perfection, which we can be guided daily, if we propose in our heart..

The truth make us free, and make the main clear and ready for a good human development in all areas that we need.

She prays for justice, peace, divine love, prosperity, be amalgamated with the peace and love of her nation and from there, be blessed to the all nations in the world.

11 GOD IS LOVE

When we defined the word love, we are talking about God, the essence of God itself.

The true love, is based in the way that God loves, as he does, and He is the source which propagates the characteristics that every human being need.

God is love. If we have God, we have love.

These four characteristics of love were created from God's heart, in the way we can developed them, having a better life according to the first source, it means, God Himself.

Now, what happened when we look around, and there is not love inside us and inside others?

If we say that we have God, but we have not love. What is happening?

We have many activities in the name of love, but there is not love on it.

"Dear friends, let us love one another, for love comes from God. Everyone who loves has been born of God and knows God. Whoever does not love does not know God, because God is love". (1 John 4: 7-8).

We can say that there is an absence of love or an

absence of God presence.

"The Holy Spirit come from Father God to possess and filled our Heart.

He, in His love, poured out the Father love in our heart.

As surely as God has given us His Spirit is true that the Holy Spirit poured out the love of God in us.

Why do we have seldom experience of this?

Simple, because our incredulity. We need time to get apart from the system of the world with its interests, in the way that our soul will delight in the light of the Lord, in the way that His eternal love will have possession of our heart.

If we believe in the infinite love of God, and in His divine power which He take possession of our heart, we will really know the love of God in our heart.

God wants that his children love Him with all heart and with all their strength. He knows what powerless we are, and for that reason He has given us His Spirit, who, search the deep things of God, and He has found the source of the eternal love which we can fill our heart".

Andrew Murray, The Power From The High.

12 THE INFLUENCE OF LOVE IN LIFE

"That which was from the beginning, which we have heard, which we have seen with our eyes, which we have looked at and our hands have touched—this we proclaim concerning the Word of life. The life appeared; we have seen it and testify to it, and we proclaim to you the eternal life, which was with the Father and has appeared to us. We proclaim to you what we have seen and heard, so that you also may have fellowship with us. And our fellowship is with the Father and with his Son, Jesus Christ. We write this to make our joy complete" (John 1-4).

When the love of God is in our life, we can see the difference and how it affect our relationship in a positive way.

The word of God said, that **the word is life, and that life living with love, is to complete our happiness and joy.**

As the same way, our energy is renewing every day, bringing to us new life and new strength.

"The Joy of the Lord is my Strength", the psalmist said.

When we have love and joy in our home, we can say that we have the necessary tools to be happy.

"And hope does not put us to shame, because God's love has been poured out into our hearts through the Holy Spirit, who has been given to us" (Romans 5:5).

What is love?

The love find joy in given everything to the person who loved to make him/her happy. And the Father God, who is in you in the deepest place of your inner sanctuary, has the goal to fill your heart with His love.

All other qualities of God have His major expression in His love.

The true and full blessing of your inner sanctuary is any less that a life submerged in the abundant of His divine love.

Hence, while you are praying, you need to believe in the love of God.

Therefore, your first and principal thought before of the Throne of God should be the faith in His love.

Have a time to have a quiet moment in the revelation of the love of Christ, until you be filled of the Spirit of worship, admiration and yearning.

If you desire this, get close to God, and remain in Him in a veneration and silent worship. Then you will know the love of God in Christ, which surpasses al understanding (Ephesians 3:19).

Andrew Murray, The Throne Of Grace

13 GOD IS LIGHT

God's light is one of the most important thing in our life.

His light is the guide of the life. If we have not His light, we have a life in darkness.

"Your word is a lamp for my feet, a light on my path" (Psalm 119:105).

"This is the message we have heard from him and declare to you: God is light; in him there is no darkness at all. If we claim to have fellowship with him and yet walk in the darkness, we lie and do not live out the truth. But if we walk in the light, as he is in the light, we have fellowship with one another, and the blood of Jesus, his Son, purifies us from all sin.

This is the message we have heard from him and proclaim to you, that God is light, and in him is no darkness at all. If we say we have fellowship with him while we walk in darkness, we lie and do not practice the truth. But if we walk in the light, as he is in the light, we have fellowship with one another, and the blood of Jesus his Son cleanses us from all sin. If we say we have no sin, we deceive ourselves, and the truth is not in us. If we confess our sins, he is faithful and just to forgive us our sins and to cleanse us from all unrighteousness. If we say we have not sinned, we make him a liar, and his word is not in us (1 John 1:5-10).

When the light of the Lord is on our heart, our lives

have His life and love.

Without the light that He give us every day, we will walk under daily shadows that the world present us every day.

With those shadows come our failed, sickness, sadness, etc. but when the light appears all that are under those shadows vanishes.

The divine light that is in God, is the light that we need as people with living life. That shine that is in our inner self, is the light that come out bringing the light of God to others.

There are people in this world that have never experimented the light of God, they are in darkness, and they do not know that they are in darkness. They believe that their way is natural, as just the way it is, but when they are illuminated by the God's light they say "WOW" I have never known this before, and then, they start to live the "living life", because they have the new light and birth.

"In the same way, let your light shine before others, that they may see your good deeds and glorify your Father in heaven" (Matthew 5:16).

The light of God is born in the heart. When we open our heart to God, the divine light come in our lives.

"Every good and perfect gift is from above, coming down from the Father of the heavenly lights, who does not change like shifting shadows" (James 1:17).

"You may have Heard say often that is someone wants a great work, you must put it all your heart and all his/her power. In temporal matters this is the secret of success and victory. And above all in divine things is something indispensable, especially when we pray asking the Holy Spirit. God grandly desires our love. For all love nature longs to be accepted and find answer. If God longs to unquenchable desire and fervent, reaching our love, the love of our all heart".

CONSUMMATED IS

By faith we can appropriate this work of the Lord on the cross, and find strength for living daily, making dying to our ego with the fellowship of the crucified Jesus.

Children of God, study the unfailing treasure that we find is this affirmation "Consummate is".

The faith that took Jesus to the cross will allow us to express, in our daily life the Spirit of the Cross.

"Grant me the grace Lord, to accomplished the things what you want me to do.

May my love do no retain anything; and lead me to surrender to your love completely Lord, Amen".

Andrew Murray, The Throne of the Grace.

14 Divine Identity

In the process of our lives, God go with us, including in our fails. He is always there, to help us, given us a way out from our problems, and He will make a better way for us, if we listen him understanding His will for us.

"**As for you, you were dead in your transgressions and sins, in which you used to live when you followed the ways of this world and of the ruler of the kingdom of the air, the spirit who is now at work in those who are disobedient. All of us also lived among them at one time, gratifying the cravings of our flesh and following its desires and thoughts. Like the rest, we were by nature deserving of wrath.**

But because of his great love for us, God, who is rich in mercy, made us alive with Christ even when we were dead in transgressions—it is by grace you have been saved. And God raised us up with Christ and seated us with him in the heavenly realms in Christ Jesus". (Ephesians 2:1-6).

The grace and love of God guide us to His presence and He just not only guide us to His presence, but He made us to be with him.

He is a compassionate God who loves his children and is ready to pour out his blessings to whom believe and hope in Him.

"**In order that in the coming ages he might show the incomparable riches of his grace, expressed in his kindness to us in Christ Jesus. For it is by grace you** have been saved, through faith—and this is not from yourselves, it is the gift of God— not by works, so that no one can boast. For we are God's handiwork, created in Christ Jesus to do good works, which God prepared in advance for us to do" (Ephesians 2: 1-1).

"As a prisoner for the Lord, then, I urge you to live a life worthy of the calling you have received. Be completely humble and gentle; be patient, bearing with one another in love. Make every effort to keep the unity of the Spirit through the bond of peace. There is one body and one Spirit, just as you were called to one hope when you were called; one Lord, one faith, one baptism; one God and Father of all, who is over all and through all and in all (Ephesians 4:1-6).

The word of God call us to a spiritual unity, but not to a religious unity or actual systematization.

The word of God said one spirit, one faith and one God.

Through Him to maintain the peace.

The world show us different system, different religions, different faith, different gods, and a continuing string of wards and hostility.

The divine original, ask and teach us to be in a spiritual unity according God's regulations.

Jesus replied, "You are in error because you do not know the Scriptures or the power of God (Matthew 22:29).

We have inheritance our ancestor's problems, spiritual, and secular education.

But, we have also inheritance the divine instructions, to have some harmony in our lives.

The law and the grace have come together to make these real, according to Jesus word.

"One of them, an expert in the law, tested him with this question: "Teacher, which is the greatest commandment in the Law?"

Jesus replied: "'Love the Lord your God with all your heart and with all your soul and with all your mind. This is the first and greatest commandment. And the second is like it: 'Love your neighbor as yourself. All the Law and the Prophets hang on these two commandments (Matthew 22: 35:39).

It seems to be a fantasy or a fairy tale, if we do not know the scriptures.

Meanwhile, the word direct us to a unity, the systems and religion divide us in colors, social status, education and culture.

Divisions bring lake of peace and joy. To be clearer, they look for wars, rivalries, evil competition etc., because these produce sicknesses which produce money, profiting with the human being.

These will be the goals for those divisions, the profit that they have as a result of these.

I can understand why Jesus was so angry with those people who were shopping at the temple.

"Jesus entered the temple courts and drove out all who were buying and selling there. He overturned the tables of the money changers and the benches of those selling doves. "It is written," he said to them, "My house will be called a house of prayer,–but you are making it a den of robbers" (Matthew 21:12-13).

The Lord said that the world will know the Christian people for the unity and love that they have each other through the Holy Spirit.

"By this everyone will know that you are my disciples, if you love one another" (John 13:35).

The blind and the lame came to him at the temple, and he healed them. But when the chief priests and the teachers of the law saw the wonderful things he did and the children shouting in the temple courts, "Hosanna to the Son of David," they were indignant.

"Do you hear what these children are saying?" they asked him.

"Yes," replied Jesus, "have you never read,

"'From the lips of children and infants you, Lord, have called forth your praise?" (Mathew 21:12-16).

45

The Jesus teachings are very actual just they were before.

If we take a look around, we can see the same attitudes, same people, and same spirits.

"Be alert and of sober mind. Your enemy the devil prowls around like a roaring lion looking for someone to devour. Resist him, standing firm in the faith, because you know that the family of believers throughout the world is undergoing the same kind of sufferings (1 Peter 5: 8-9).

It is our responsibility to know and develop the divine instructions which are our heritance though the promises and sacrifice of Jesus on the Cross, having trusting and faith, we need to confront the circumstances every day.

"For he himself is our peace, who has made the two groups one and has destroyed the barrier, the dividing wall of hostility, by setting aside in his flesh the law with its commands and regulations. His purpose was to create in himself one new humanity out of the two, thus making peace, and in one body to reconcile both of them to God through the cross, by which he put to death their hostility. He came and preached peace to you who were far away and peace to those who were near. For through him we both have access to the Father by one Spirit" (Ephesians 2: 14-18).

This is the confidence we have in approaching God: that if we ask anything according to his will, he hears us (1 John 5:14).

We have a God in whom we can trust completely.

His fidelity goes beyond our achievements and merits. His big love is the engine of our salvation.

"Let us then approach God's throne of grace with confidence, so that we may receive mercy and find grace to help us in our time of need" (Hebrews 4:16).

We can approach God in any time we need, just be careful to be honest in spirit and true, and the way that we no spent our time trying to give a show to God, because he really knows our hearts.

"You know when I sit and when I rise; You perceive my thoughts from afar, You discern my going out and my lying down, you are familiar in ALL MY WAYS, before a word in on my tongue you, Lord, knows it completely (Psalms 139: 2-4).

God is pleased in take care and guiding his children, protecting and guiding them in justice, love and true.

THE INGREDIENT OF PRAYERS

"**And I will do whatever you ask in my name, so that the Father may be glorified in the Son**" (John 14:13).

In this technology time that we are living in, we have all what we need to be in communication each other.

When we have the time to talk and share our lives with other is when our lives and experiences are growing.

At the same way, when we have our time with the Father God, and have the conversation with Him, He poured out His experiences and word through the Holy Spirit on us, and our inner self grows in wisdom and faith.

When the communication that we have with The Father God and us is involved with the Spirit of Love and Hope, that prayers are like the air that we breathe daily.

While you are praying, you need to believe in the love of God, the love that He has poured out through the Holy Spirit (Romans 5:5).

Prayer has given us in the way that we can communicate to God.

God Has given us His word in the way we can understand His heart language.

When we know a person, we know how to talk with them.

If we know God's word, we will know how to talk with him through his word and Holy Spirit.

"Jesus said: **And I will do whatever you ask in my name".**

He will be interceding in our prayers, in the way that the Father be glorified in the Son.

Our God yearn for that time with us.

He is a God who is pleased to fill any desire of our heart.

He love the time that He share with their children.

He love the worship time with us.

We have the way to make our lives be full of his love, wisdom and grace.

If we desires that our inner sanctuary flows with his love and wisdom, we need to spend time with him, in the way He can guide us to his throne of grace.

"I will give you the keys of the kingdom of heaven; whatever you bind on earth will be bound in heaven, whatever you unbid on earth will be unbid in heaven" (Matthew 16:19).

We can bind the blessings that he has for us in heaven and we can bind the curses in our lives on the earth, through the prayers, cover them with the blood of Jesus.

"We need to understand the importance of the first meeting of praying for the history of the kingdom.

That meeting is that give us the key of an continuous presence of Christ and the power of His Spirit; the key that open the barn door of heaven with all blessings.

It is the answer of the prayers in unity when the heaven was opened and the Spirit of God descended to the earth to dwell in their hearts".

Andrew Murray, The Power Of The Praying.

About The Author

I am Elizabeth Escauriza Hempel mother of four children, and I have an immeasurable love for God.

I have written various little books as With All Heart, How To Speak In Public, In Your Presence, The Parables of Jesus, The Complete Name Of The Love, Divine Identity, The Fruit and Values Of Life, The Holy Spirit Image, The Divine Software, Reaching the Blessings Of God Though The Prayers, among others.

It is my desire to share the word of God and I hope these books be a blessing to your lives.

"Holy Spirit concede me that in my spiritual life have the real experience of knowing and experiment the powerful work of The Holy Spirit in every and each day of my days".

Andrew Murray, The Power From The High.

www.ingramcontent.com/pod-product-compliance
Lightning Source LLC
Chambersburg PA
CBHW051821170526
45167CB00005B/2105